KA MOʻOLELO HELU

THE NUMBER STORY

SMALL BOOK ONE

ENGLISH - HAWAIIAN

*Numbers Teach Children
Their Number Names*

written and illustrated by

MISS ANNA

Early Reader Edition of *The Number Story 1*
Bronze Medal Winner, 2016 Wishing Shelf Book Award

Cover by | Lumpy Publishing
Layout by | Lumpy Publishing
Translated by Poiboi
Coloring by Jieeun Woo and Maria Mirabella

Library of Congress Control Number: 2018902040

Names: Miss Anna, author.
Title: Number story : numbers teach children their number names / Miss Anna.
Description: Portland, OR: Lumpy Publishing, 2018.
Identifiers: ISBN 978-1-949320-24-4 | LCCN 2018902040
Summary: The pictures and rhymes present stories which introduce numbers 0-10.
Subjects: LCSH Numeration—English—Hawaiian-Pictorial works--Juvenile literature. | BISAC JUVENILE NONFICTION /
Languages: English—Hawaiian
Classification: LCC QA141.3 .M57 2018 | DDC 513—dc23

Publisher: Lumpy Publishing
Website: www.missannabooks.com
Email: missanna@missannabooks.com

Paperback: ISBN 978-1-949320-24-4
Printed in the U.S.A.   1 3 5 7 9 10 8 6 4 2

Makemake ʻoe e
hoʻopaʻa i ka helu inoa?

It is very easy and a lot of fun!

He maʻalahi loa kēia a me le'aleʻa loa!

Say-along our little jingle

E hīmeni ia mākou i kēia moʻolelo liʻiliʻi!

starting from Number One!

E hoʻomaka ana ia mākou mai helu ʻekahi!

1

ONE looks like my one finger.

ʻEKAHI

like me koʻu manamana kuhi.

ONE!
'EKAHI!

2

TWO trails a tail.

ʻELUA

ke hoʻokolo nei i ka huelo.

A TAIL! HE HUELO!

3

THREE has bumps.

'EKOLU

he 'anapu'u ko'u.

BUMPY! 'ANAPU'U!

FOUR carries a sail.

'EHĀ

ke hāʻawe nei i ka lā.

A SAIL!
HE LĀ!

5

FIVE   is a racing track.

'ELIMA

he kahua heihei kēia.

VROOM
VROOOM!

6

SIX curves like a snail.

ʻEONO

ke piʻo nei like me ke kamaloli.

HE KAMALOLI!

7

SEVEN has a sharp angle.

'EHIKU

he huina kūpono wini ko'u.

BE CAREFUL! IT'S SHARP!

AKAHELE! HE WINI KĒIA!

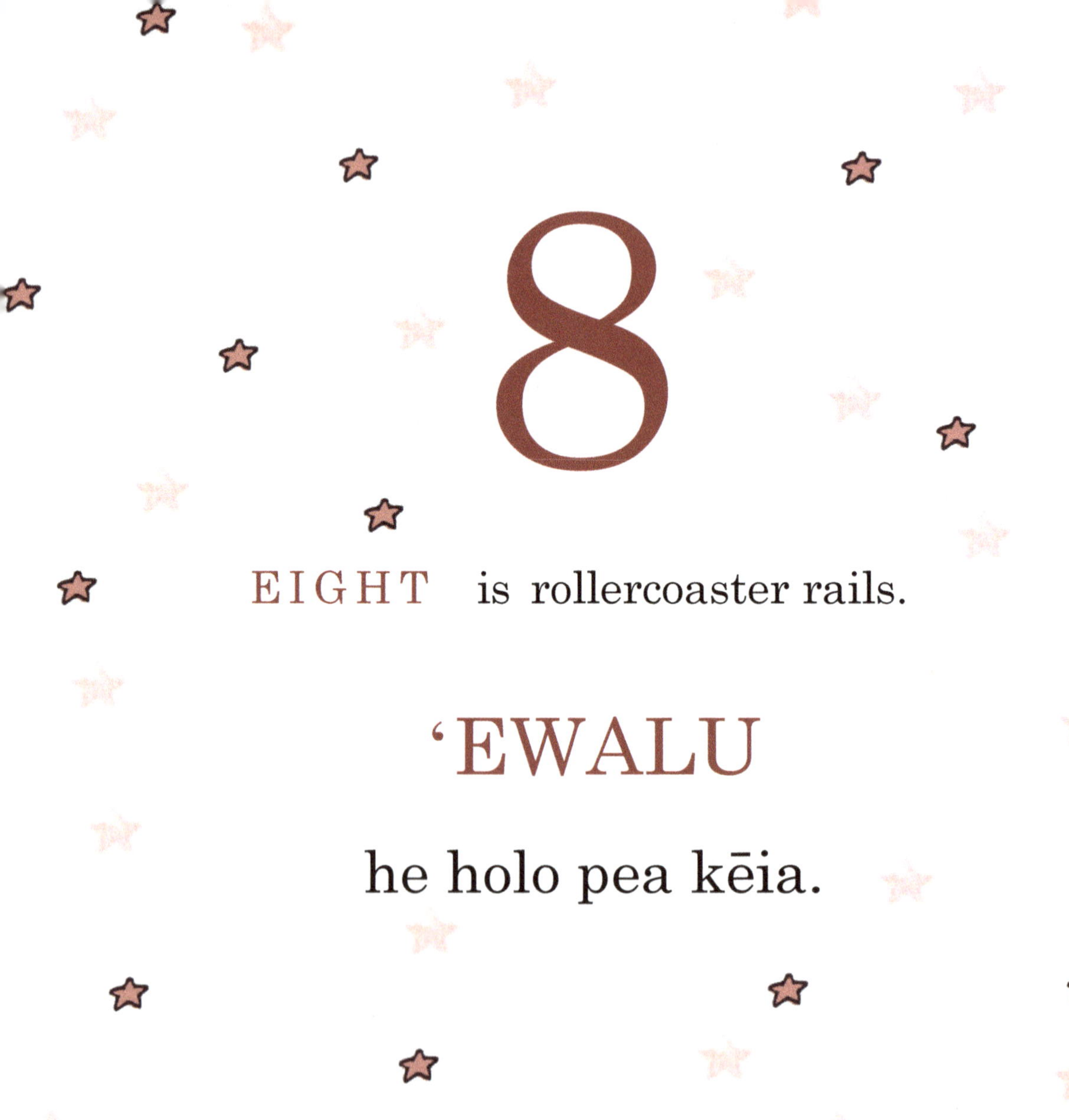
8

EIGHT   is rollercoaster rails.

'EWALU

he holo pea kēia.

HULO!
YIPPEE!

9

NINE is a bubble on a stick.

'EIWA

he hu'a kopa ma ka mea
omo kēia.

A BUBBLE!

HE HUʻA KOPA!

10

TEN is an eye of a whale.

'UMI

he maka o ke koholā kēia.

WINK!

'IMO!

HELLO! ALOHA!

And
A me
0
ZERO   is an empty pail.
'OLE
he kini pela ka'ele kēia.

IT'S
EMPTY!
KA'ELE!

Thank you for playing with us today.

We had a lot of fun too!

Mahalo ia ʻoe no pāʻani me mākou i kēia lā.

Ua leʻaleʻa ʻo au mākou!

We are your Number friends,
Zero to Ten,
Who will be here for you~
He helu na hoaloha e mākou
'ole iā 'umi.
Aloha nui wau ia 'oe.

Bye-bye now!
See you again soon!
A hui hou!
E 'ike hou auane'i!

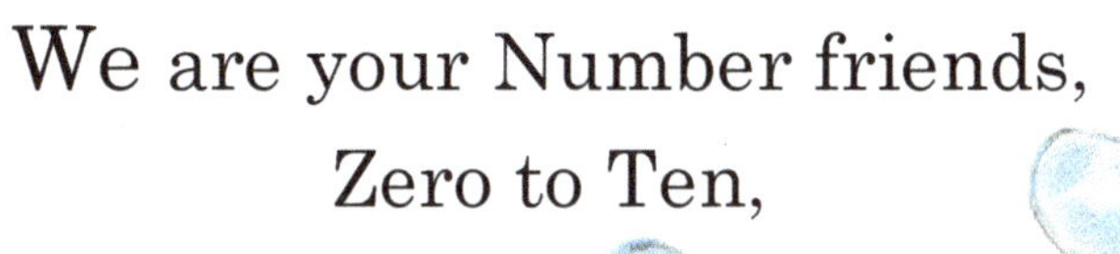

The Numbers are *SINGING* too!

To sing-a-long, look for Miss Anna Number Story
at your favorite music store like iTUNES.

MP3

Numbers 0-10
IDENTIFYING & COUNTING

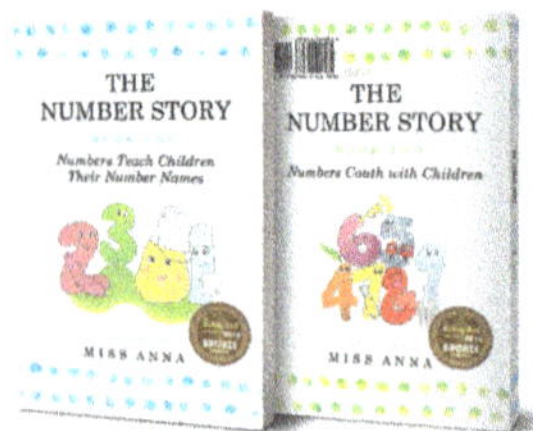

Number Story 1 & 2

isbn: 978-0-996216-48-7

Numbers 11-20 & Ordinals
first, second, third...

Number Story 3 & 4

isbn: 978-1-945977-01-5

Numbers 0-100 & Place Values
ones, tens, hundreds...

Number Story 5 & 6

isbn: 978-1-945977-06-0

About Clocks & Telling Time
hours, minutes, seconds

Number Story 7 & 8

isbn: 978-1-949320-40-4

For more Miss Anna books to love,
visit us at

www.missannabooks.com

Numbers are working hard all over the world!
Come Travel the World with Us!

* 9 7 8 1 9 4 9 3 2 0 2 4 4 *